AHWOOOOO! The wolf's call rings through the air. It can be heard for miles around.

How would you feel if you heard this call? Many people believe that a wolf is an evil animal like the scary ones portrayed in stories. But these stories are not true. Do you know the real story about wolves?

Wolves have roamed the Earth for over a million years. The most common kind of wolf is *Canis lupus* (CAN-is LOOP-us), or **gray wolf.** But not all gray wolves are gray.

No two gray wolves look exactly alike. They can be any color—ranging from white to black—and are different sizes and weights.

Gray wolves from different habitats are often called by other names. Wolves that live in forests are called **timber wolves**. And **tundra wolves** or **arctic wolves** are the names for wolves that live in cold, arctic regions.

Timber wolves can be black, dark brown, or dark gray. Their coloring makes it hard for them to be seen among the trees.

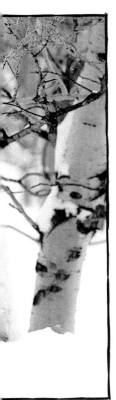

The tundra or arctic wolf's long, thick fur keeps it warm and protects it against the bitter-cold winds and ice. Pure white fur helps this wolf blend in to its snowy surroundings.

Like people, wolves live in families, or **packs.**
The mother and father are the leaders, called
alpha wolves. And they dominate, or rule, the
pack. Each of the other members has its own
place, or rank, below the pack leaders.

Packs—as small as
three or as large as
thirty—are made up of
a father and mother,
their young, and a few
other relatives.

Each pack has its own
living and hunting area,
called a **territory.** The
pack leader marks the
territory by leaving scent
marks. This warns other
packs to stay away.

In play or in a fight, pack members may snarl and bare their teeth. But, before there is bloodshed, the lower-ranking wolf rolls over on its back to show that it gives up.

The least powerful male and female pack members are called *omega* wolves. Sometimes an omega wolf is forced out of the pack. Now a "lone" wolf, it must travel long distances to find food.

Pups are usually born during the spring. They weigh about a pound and cannot see or hear. Newborn pups feed on milk they get from their mother. They are protected by the pack and are hidden in an underground den.

There are usually five or six pups in a litter. At three weeks old the pups are able to go outside, but still sleep in the den, where they are safe and can be easily watched.

When the pups get older, an adult from the pack carries food to them. The food is sometimes carried in the adult's mouth, but usually it's swallowed and carried in the stomach. This prevents other animals from getting the food. When the adult reaches the pups, it spits up the food, or *regurgitates* (ree-GIR-ji-TATES), into each pup's mouth.

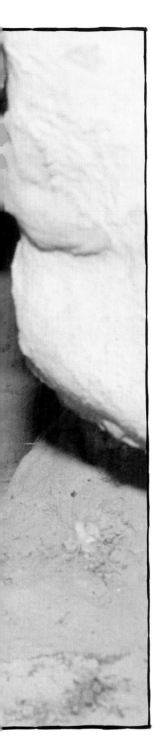

Month-old pups tumble and wrestle among themselves. This is partly play, but it also determines which male and which female are the *betas*, or strongest puppies. The betas only dominate other puppies, but as time goes by and the betas become adults, they may one day lead the pack.

All pack members take care of the pups. At least one wolf is left to babysit the pups while the rest of the pack looks for food. This "babysitter" is taking the pups from the den to a place nearby where they can play.

The pups' playful tugging is one way for them to learn how to use their strong jaws and sharp teeth. This is practice for the hunt. When the pups are full-grown, they will have to know how to catch, hold, and eat their prey.

All animals eat in order to survive. Wolves must hunt other animals for food. It sometimes takes a wolf pack several days to catch its prey. The pack depends mostly on larger animals, like moose, deer, elk, mountain sheep, and musk oxen, for food.

Now the wolf pack chases the herd. The herd starts to run. Soon, the pack will strike.

The pack studies the herd of musk oxen to find a weak animal, one that may not be able to defend itself.

The pack surrounds the selected animal and will bring it down together.

After the leaders eat, the others get their turn. Each wolf can eat as much as 20 pounds of meat.

What makes the wolf such an able hunter? Its sense of smell and hearing combined with its size, speed, and intelligence. Scientists believe that the wolf is highly intelligent—some say, it is even smarter than the dog.

The wolf is able to run long distances. Strong muscles and long legs carry it quickly over very rough terrain.

A wolf's ears can move in different directions to find sound. It points its ears in the direction where the sound is coming from. A wolf can hear high-pitched sounds and it can hear another wolf howling six to seven miles away.

Some scientists believe a wolf's sense of smell is more than a thousand times better than a person's. A wolf's nose is so sensitive that it can smell its prey from more than a mile away.

Wolves use their voices to communicate. Scientists that study animals have observed that a pack will howl before and after a hunt. The reason may be to tell other packs to stay away.

Each wolf's voice is unique. When more than one wolf howls, it sounds almost like singing.

A wolf may also howl when it is lost, or wants the pack to gather together. Howling may express joy or sadness. Other sounds the wolf makes are growling, yipping, and barking. Here, a pup is practicing its howl.

The lower-ranking wolf lowers its head and licks the face of the higher-ranking pack member. This behavior communicates respect.

Wolves are also able to express themselves with their faces and their bodies. What do you think these wolves are communicating?

The wolves' tails are up and their bodies are relaxed. They are playing. Some scientists believe you can see the wolves' happy expressions on their faces when they play. Can you?

Begging is how pups tell the adults they are hungry.

Snarling, baring teeth, and a tense, crouched body with the tail down show anger.

When people cleared forests to build farms and ranches, wolves lost their natural habitats. Some wolves came into settled areas looking for food. Afraid for their livestock—horses, cattle, and sheep—many people started to kill the wolves. Scientists went looking for the wolves that were left to try to protect and study them. But very few wolves were found.

Here, a scientist has placed a radio transmitter collar around the wolf's neck. This will allow the scientist to track the wolf's movements. The scientist hopes the wolf will lead him to other wolves or packs.

Today, in almost every part of the United States, wolves are an extinct species. But Canada still has many wolves. This gave some scientists an idea—to *reintroduce,* or bring wolves from Canada back to the United States.

After a Canadian wolf is captured, scientists record its weight, length, and coloring. To monitor the wolf's movements once it is set free, the wolf will be tagged and fitted with a radio transmitter collar.

The wolf is then placed in a cage so it can be safely transported to a wilderness site in the United States.

Finally, at a site in Yellowstone National Park, the wolf is released from its cage to live in the wild.

Now you know the real story about wolves. They are not evil and cruel, but highly intelligent animals with a family structure very much like our own. Wolves were once a natural part of our environment. Today there is hope that they will be again.